School Chemistry Laboratory Safety Guide

October 2006

U.S. Consumer Safety Product Commission

DEPARTMENT OF HEALTH AND HUMAN SERVICES
Centers for Disease Control and Prevention
National Institute for Occupational Safety and Health

The views or opinions expressed in this safety guide do not necessarily represent the views of the Commission.

> This document is in the public domain and may be freely copied or reprinted.

Disclaimer

Mention of the name of any company or product does not constitute endorsement by the U.S. Consumer Product Safety Commission (CPSC) and the National Institute for Occupational Safety and Health (NIOSH). In addition, citations to Web sites do not constitute CPSC and NIOSH endorsement of the sponsoring organizations or their programs or products. Furthermore, CPSC and NIOSH are not responsible for the content of these Web sites.

Ordering Information

CPSC

Access through the Internet

This guide along with other CPSC news releases, Public Calendar and other information can be obtained via the Internet from the agency's Web site at **www.cpsc.gov**

[For ordering hard copies of publications and publications: publications@cpsc.gov. Please allow 3–4 weeks for delivery.]

or write to
U.S. Consumer Product Safety Commission
4330 East West Highway
Bethesda, MD 20814

CPSC Consumer Hotline
English/Spanish: 1–800–638–2772
Hearing/Speech Impaired: 1–800–638–8270

CPSC Publication No. 390

NIOSH

To receive documents or other information about occupational safety and health topics, contact NIOSH at:

NIOSH—Publications Dissemination
4676 Columbia Parkway
Cincinnati, OH 45226–1998

Telephone: **1–800–35–NIOSH** (1–800–356–4674)
Fax: 513–533–8573
E-mail: pubstaft@cdc.gov

or visit the NIOSH Web site at **www.cdc.gov/niosh**

DHHS (NIOSH) Publication No. 2007–107

Pull-Outs

The following pages are available at the end of the document for easy copying for distribution or posting:

Safety Do's and Don'ts for Students
How Should Chemicals Be Stored?
Suggested Shelf Storage Pattern

Foreword

In 1984, the Council of State Science Supervisors, in association with the U.S. Consumer Product Safety Commission and the National Institute for Occupational Safety and Health, published the safety guide *School Science Laboratories: A Guide to Some Hazardous Substances* to help science teachers identify hazardous substances that may be used in school laboratories and provide an inventory of these substances.

Because school science curricula have changed since then, the safety guide has been updated and revised to reflect those changes. This guide on safety in the chemistry laboratory was also written to provide high school chemistry teachers with an easy-to-read reference to create a safe learning environment in the laboratory for their students. The document attempts to provide teachers, and ultimately their students, with information so that they can take the appropriate precautionary actions in order to prevent or minimize hazards, harmful exposures, and injuries in the laboratory.

The guide presents information about ordering, using, storing, and maintaining chemicals in the high school laboratory. The guide also provides information about chemical waste, safety and emergency equipment, assessing chemical hazards, common safety symbols and signs, and fundamental resources relating to chemical safety, such as Material Safety Data Sheets and Chemical Hygiene Plans, to help create a safe environment for learning. In addition, checklists are provided for both teachers and students that highlight important information for working in the laboratory and identify hazards and safe work procedures.

This guide is not intended to address all safety issues, but rather to provide basic information about important components of safety in the chemistry laboratory and to serve as a resource to locate further information.

Nancy A. Nord
Acting Chairman, U.S. Consumer
   Product Safety Commission

John Howard, M.D.
Director, National Institute for Occupational
   Safety and Health
   Centers for Disease Control and Prevention

Contents

Disclaimer ii
Foreward v
Acknowledgments viii

Introduction 1
What are the Teacher's Responsibilities? 3
What are the Safety Do's and Don'ts for Students? 6
What is a Chemical Hygiene Plan? 10
What is a Material Safety Data Sheet? 12
What Should be Considered When Purchasing Chemicals? 13
What is a Chemical Tracking System and How Should It be Set Up? ... 15
How Should Chemical Containers be Labeled? 17
How Should Chemicals be Stored? 19
Suggested Shelf Storage Pattern 21
Suggested Shelf Storage Pattern for Inorganics 22
Suggested Shelf Storage Pattern for Organics 23
How Should Compressed Gas Cylinders be Stored, Maintained,
   and Handled? 24
What are Some Strategies to Reduce the Amount and/or
Toxicity of Chemical Waste Generated in the Laboratory? 26
What is the Recommended Procedure for Chemical Disposal? 28

Appendices
   A. Common Safety Symbols 30
   B. National Fire Protection Association Hazard Labels 31
   C. Substances with Greater Hazardous Nature
         Than Educational Utility 33
   D. Substances with a Hazardous Nature,
         But May Have Potential Educational Utility 40
   E. Incompatible Chemicals 44
   F. Recommended Safety and Emergency Equipment
         for the Laboratory 47
   G. How Does a Chemical Enter the Body? 48
   H. What are Exposure Limits? 50
   I. General Guidelines to Follow in the Event of a
         Chemical Accident or Spill 52
   J. Understanding an MSDS 54
   K. Sample MSDS 56
   L. Web Site Resources 63
   M. Glossary 65

Pull-outs 72

Acknowledgments

This safety guide was written, revised, and reviewed by scientists from the Consumer Product Safety Commission (CPSC), the National Institute for Occupational Safety and Health (NIOSH), and Environmental Protection Agency (EPA). Kailash Gupta, DVM, Ph.D., Directorate for Health Sciences, served as the CPSC project officer; Patricia Brundage, Ph.D., Directorate for Health Sciences, CPSC served as author, and John Palassis, CIH, CSP, CHMM, Education and Information Division, NIOSH served as the project officer and a co-author.

Lori Saltzman, MS, Mary Ann Danello, PhD, from the Directorate for Health Sciences, CPSC, Charles Geraci, Ph.D., TJ Lentz, Ph.D., Ralph Zumwalde, Alan Weinrich, Michael Ottlinger, Ph.D., from the NIOSH Education and Information Division, from the Office Of Director, NIOSH, provided critical review and input.

Staff in the Office of Public Affairs at CPSC provided editorial, design and production assistance. In NIOSH, Susan Afanuh provided editorial services, and Vanessa Becks and Gino Fazio provided desktop design and production assistance.

The safety guide was reviewed with the assistance of American Chemical Society, the National Institute of Environmental Health Sciences/National Institute of Health, the Council of State Science Supervisors, American Federation of Teachers/AFL-CIO, Cincinnati Federation of Teachers, National Science Teachers Association, Environmental Protection Agency, Federal OSHA Directorate of Standards and Guidance, Federal OSHA, Region VII.

Introduction

Recognition of laboratory safety and health problems has crystallized since the passage of the Occupational Safety and Health Act of 1970. This Act requires that certain precautions be observed to protect the safety and health of employees on the job. The employee designation includes all teachers employed by private and public school systems in States that have occupational safety and health plans accepted by the Occupational Safety and Health Administration (OSHA) of the U.S. Department of Labor (DOL). OSHA rules and regulations are provided to protect the employees and the facilities.

The importance of laboratory safety has been recognized for many years in industry. However, educational institutions have been slower to adopt such safety practices and programs.

A science program has certain potential dangers. Yet, with careful planning, most dangers can be avoided in an activity-oriented science program. It is essential for all involved in the science instruction program to develop a positive approach to a safe and healthful environment in the laboratory. Safety and the enforcement of safety regulations and laws in the science classroom and laboratory are the responsibility of the principal, teacher, and student—each assuming his/her share. Safety and health should be an integral part of the planning, preparation, and implementation of any science program.

The Importance of Safety

Safety and health considerations are as important as any other materials taught in high school science curricula. Occupational injury data from industry studies indicate that the injury rate is highest during the initial period of employment and decreases with experience. Similarly, in a high school laboratory setting where students experience new activities, the likelihood of incidents, injury, and damage is high. Therefore, it is essential that the students are taught what can go wrong, how to prevent such events from occurring, and what to do in case of an emergency.

Teacher's / Instructor's Viewpoint

Teachers have an obligation to instruct their students in the basic safety practices required in science laboratories. They also have an obligation to instruct them in the basic principles of health hazards that are found in most middle and secondary school science laboratories. Instructors must provide safety information and training to the students for every stage of experiment planning and be there to observe, supervise, instruct, and correct during the experimentation. Teachers play the most important role in insuring a safe and healthful learning environment for the students. The ideal time to impress on students' minds the need for caution and preparation is before and while they are working with chemicals in science laboratories.

Student's Viewpoint

Students develop attitudes towards safety and acquire habits of assessing hazards and risks when they are young. Students come from diverse backgrounds and have various levels of preparation. Most of them have no previous hands on training in handling chemicals or equipment; others may come well prepared to assume personal responsibility for risk assessment and safety planning in their experiments. The school science laboratory provides an opportunity to instill good attitudes and habits by allowing students to observe and select appropriate practices and perform laboratory operations safely. Safety and health training lays the foundation for acquiring these skills. The students should think through implications and risks of experiments that they observe or conduct in order to learn that safe procedures are part of the way science must be done.

Student motivation in any area of education is a critical factor in the learning process. Emphasizing the importance of safety and health considerations by devoting substantial class time to these areas should help. The current popular preoccupation with matters of industrial safety and health may also serve as motivation. Students may find a discussion of toxicology interesting, informative, and beneficial. The possibilities for working this material into the science curriculum are innumerable and limited only by the imagination of the teacher.

School's Viewpoint

Support for laboratory safety programs is the responsibility of school system administrators. School system administrators should appreciate the need for establishing safety and health instruction as a fundamental part of a science curriculum and should operate their schools in as safe a manner as possible.

No Federal law requires safety and health programs to protect students in schools. The Occupational Safety and Health Act of 1970 requires employers to provide safety and health protection for teachers and other school system employees. Some States (North Carolina, for example) require school systems to abide by State regulations, which are similar to the OSHA Laboratory Standard (29 CFR 1910.1450).

All safety programs must actively involve the school administrators, supervisors, teachers, and students, and all have the responsibility for safety and health of every other person in the laboratory and school.

What are the Teacher's Responsibilities?

Teachers and teacher-aides should lead by example and wear personal protective equipment; follow and enforce safety rules, procedures, and practices; and demonstrate safety behavior and promote a culture of safety. They should be proactive in every aspect of laboratory safety, making safety a priority. The following is a checklist for teachers highlighting essential information for working in the high school laboratory. This is a general safety checklist and should be periodically re-evaluated for updates.

Upkeep of Laboratory and Equipment

- Conduct regular inspections of safety and first aid equipment as often as requested by the administration. Record the inspection date and the inspector's initials on the attached equipment inspection tag.
- Notify the administration in writing if a hazardous or possibly hazardous condition (e.g., malfunctioning safety equipment or chemical hazard) is identified in the laboratory and follow through on the status.
- Never use defective equipment.

Recordkeeping

- Keep organized records on safety training of staff for as long as required by the school system.
- Keep records of all laboratory incidents for as long as required by the school system.

Safety and Emergency Procedures

- Educate students on the location and use of all safety and emergency equipment prior to laboratory activity.
- Identify safety procedures to follow in the event of an emergency/accident.
- Provide students with verbal and written safety procedures to follow in the event of an emergency/accident.
- Know the location of and how to use the cut-off switches and valves for the water, gas, and electricity in the laboratory.

What are the Teacher's Responsibilities?

- Know the location of and how to use all safety and emergency equipment (i.e., safety shower, eyewash, first-aid kit, fire blanket, fire extinguishers and mercury spill kits).
- Keep a list of emergency phone numbers near the phone.
- Conduct appropriate safety and evacuation drills on a regular basis.
- Explain in detail to students the consequences of violating safety rules and procedures.

Maintenance of Chemicals

- Perform regular inventory inspections of chemicals.
- Update the chemical inventory at least annually, or as requested by the administration.
- Provide a copy of the chemical inventory to the local emergency responders (i.e., fire department).
- Do not store food and drink with any chemicals.
- If possible, keep all chemicals in their original containers.
- Make sure all chemicals and reagents are labeled.
- Do not store chemicals on the lab bench, on the floor, or in the laboratory chemical hood.
- Ensure chemicals not in use are stored in a locked facility with limited access.
- Know the storage, handling, and disposal requirements for each chemical used.
- Make certain chemicals are disposed of properly. Consult the label and the Material Safety Data Sheet for disposal information and always follow appropriate chemical disposal regulations.

Preparing for Laboratory Activities

- Before each activity in the laboratory, weigh the potential risk factors against the educational value.
- Have an understanding of all the potential hazards of the materials, the process, and the equipment involved in every laboratory activity.
- Inspect all equipment/apparatus in the laboratory before use.
- Before entering the laboratory, instruct students on all laboratory procedures that will be conducted.
- Discuss all safety concerns and potential hazards related to the laboratory work that students will be performing before starting the work. Document in lesson plan book.

Ensuring Appropriate Laboratory Conduct

- Be a model for good safety conduct for students to follow.
- Make sure students are wearing the appropriate personal protective equipment (i.e., chemical splash goggles, laboratory aprons or coats, and gloves).
- Enforce all safety rules and procedures at all times.
- Never leave students unsupervised in the laboratory.
- Never allow unauthorized visitors to enter the laboratory.
- Never allow students to take chemicals out of the laboratory.
- Never permit smoking, food, beverages, or gum in the laboratory.

What are the Safety Do's and Don'ts for Students?

Life threatening injuries can happen in the laboratory. For that reason, students need to be informed of the correct way to act and things to do in the laboratory. The following is a safety checklist that can be used as a handout to students to acquaint them with the safety do's and don'ts in the laboratory.

Conduct

- Do not engage in practical jokes or boisterous conduct in the laboratory.
- Never run in the laboratory.
- The use of personal audio or video equipment is prohibited in the laboratory.
- The performance of unauthorized experiments is strictly forbidden.
- Do not sit on laboratory benches.

General Work Procedure

- Know emergency procedures.
- Never work in the laboratory without the supervision of a teacher.
- Always perform the experiments or work precisely as directed by the teacher.
- Immediately report any spills, accidents, or injuries to a teacher.
- Never leave experiments while in progress.
- Never attempt to catch a falling object.
- Be careful when handling hot glassware and apparatus in the laboratory. Hot glassware looks just like cold glassware.
- Never point the open end of a test tube containing a substance at yourself or others.
- Never fill a pipette using mouth suction. Always use a pipetting device.
- Make sure no flammable solvents are in the surrounding area when lighting a flame.
- Do not leave lit Bunsen burners unattended.
- Turn off all heating apparatus, gas valves, and water faucets when not in use.
- Do not remove any equipment or chemicals from the laboratory.

- Coats, bags, and other personal items must be stored in designated areas, not on the bench tops or in the aisle ways.
- Notify your teacher of any sensitivities that you may have to particular chemicals if known.
- Keep the floor clear of all objects (e.g., ice, small objects, spilled liquids).

Housekeeping

- Keep work area neat and free of any unnecessary objects.
- Thoroughly clean your laboratory work space at the end of the laboratory session.
- Do not block the sink drains with debris.
- Never block access to exits or emergency equipment.
- Inspect all equipment for damage (cracks, defects, etc.) prior to use; do not use damaged equipment.
- Never pour chemical waste into the sink drains or wastebaskets.
- Place chemical waste in appropriately labeled waste containers.
- Properly dispose of broken glassware and other sharp objects (e.g., syringe needles) immediately in designated containers.
- Properly dispose of weigh boats, gloves, filter paper, and paper towels in the laboratory.

Apparel in the Laboratory

- Always wear appropriate eye protection (i.e., chemical splash goggles) in the laboratory.
- Wear disposable gloves, as provided in the laboratory, when handling hazardous materials. Remove the gloves before exiting the laboratory.
- Wear a full-length, long-sleeved laboratory coat or chemical-resistant apron.
- Wear shoes that adequately cover the whole foot; low-heeled shoes with non-slip soles are preferable. Do not wear sandals, open-toed shoes, open-backed shoes, or high-heeled shoes in the laboratory.
- Avoid wearing shirts exposing the torso, shorts, or short skirts; long pants that completely cover the legs are preferable.
- Secure long hair and loose clothing (especially loose long sleeves, neck ties, or scarves).
- Remove jewelry (especially dangling jewelry).

- Synthetic finger nails are not recommended in the laboratory; they are made of extremely flammable polymers which can burn to completion and are not easily extinguished.

Hygiene Practices

- Keep your hands away from your face, eyes, mouth, and body while using chemicals.
- Food and drink, open or closed, should never be brought into the laboratory or chemical storage area.
- Never use laboratory glassware for eating or drinking purposes.
- Do not apply cosmetics while in the laboratory or storage area.
- Wash hands after removing gloves, and before leaving the laboratory.
- Remove any protective equipment (i.e., gloves, lab coat or apron, chemical splash goggles) before leaving the laboratory.

Emergency Procedure

- Know the location of all the exits in the laboratory and building.
- Know the location of the emergency phone.
- Know the location of and know how to operate the following:
  - Fire extinguishers
  - Alarm systems with pull stations
  - Fire blankets
  - Eye washes
  - First-aid kits
  - Deluge safety showers
- In case of an emergency or accident, follow the established emergency plan as explained by the teacher and evacuate the building via the nearest exit.

Chemical Handling

- Check the label to verify it is the correct substance before using it.
- Wear appropriate chemical resistant gloves before handling chemicals. Gloves are not universally protective against all chemicals.
- If you transfer chemicals from their original containers, label chemical containers as to the contents, concentration, hazard, date, and your initials.

What are the Safety Do's and Don'ts for Students?

- Always use a spatula or scoopula to remove a solid reagent from a container.
- Do not directly touch any chemical with your hands.
- Never use a metal spatula when working with peroxides. Metals will decompose explosively with peroxides.
- Hold containers away from the body when transferring a chemical or solution from one container to another.
- Use a hot water bath to heat flammable liquids. Never heat directly with a flame.
- Add concentrated acid to water slowly. Never add water to a concentrated acid.
- Weigh out or remove only the amount of chemical you will need. Do not return the excess to its original container, but properly dispose of it in the appropriate waste container.
- Never touch, taste, or smell any reagents.
- Never place the container directly under your nose and inhale the vapors.
- Never mix or use chemicals not called for in the laboratory exercise.
- Use the laboratory chemical hood, if available, when there is a possibility of release of toxic chemical vapors, dust, or gases. When using a hood, the sash opening should be kept at a minimum to protect the user and to ensure efficient operation of the hood. Keep your head and body outside of the hood face. Chemicals and equipment should be placed at least six inches within the hood to ensure proper air flow.
- Clean up all spills properly and promptly as instructed by the teacher.
- Dispose of chemicals as instructed by the teacher.
- When transporting chemicals (especially 250 mL or more), place the immediate container in a secondary container or bucket (rubber, metal or plastic) designed to be carried and large enough to hold the entire contents of the chemical.
- Never handle bottles that are wet or too heavy for you.
- Use equipment (glassware, Bunsen burner, etc.) in the correct way, as indicated by the teacher.

What is a Chemical Hygiene Plan?

A chemical hygiene plan (CHP) is a written program stating the policies, procedures, and responsibilities that serve to protect employees from the health hazards associated with the hazardous chemicals used in that particular workplace.

- OSHA's *Occupational Exposure to Hazardous Chemicals in Laboratories Standard* (Title 29, Code of Federal Regulations, Part 1910.1450, specifies the mandatory requirements of a CHP to protect persons from harm due to hazardous chemicals. The Standard can be viewed on the OSHA Web site at www.osha.gov.
- It applies to school employees who work in laboratory settings (i.e., science teachers and lab assistants); indirectly it may serve to protect students.
- The school superintendent, science department chairperson, and/or chemistry teacher(s) are typically responsible for developing the CHP for the school.
- Appendix A of 29 Code of Federal Regulations 1910.1450 provides non-mandatory recommendations to assist in the development of a CHP.

Chemical Hygiene Plan Required Elements

1. Defined standard operating procedures relevant to safety and health considerations for each activity involving the use of hazardous chemicals.
2. Criteria to use to determine and implement control measures to reduce exposure to hazardous materials (i.e., engineering controls, the use of personal protective equipment, administrative controls, and hygiene practices) with particular attention given to the selection of control measures for extremely hazardous materials.
3. A requirement to ensure laboratory chemical hoods and other protective equipment are installed and functioning properly.
4. Information for persons working with hazardous substances specifying the hazards of the chemicals in the work area, the location of the CHP, signs and symptoms associated with hazardous chemical exposures, the permissible or recommended exposure limits of the chemicals, and the location and availability of information on the hazards, safe handling, storage, and disposal of hazardous chemicals [not limited to Material Safety Data Sheets (MSDSs)].

5. Training for persons working with hazardous substances that includes methods and observations to detect the presence or release of a hazardous chemical, the physical and health hazards of the chemicals used, the measures to be taken to protect against these hazards (i.e., personal protective equipment, appropriate work practices, emergency response actions), and applicable details of the CHP.
6. The circumstances under which a particular laboratory operation or procedure requires prior approval from the appropriate administrator.
7. Requirements for medical consultation and medical examination whenever (1) a person develops signs or symptoms associated with a hazardous chemical, (2) exposure monitoring reveals an exposure level routinely above the action level, or (3) an event takes place in the work area such as a spill, leak, explosion or other occurrence resulting in the likelihood of a hazardous exposure.
8. Designation of personnel responsible for the implementation of the CHP, including the assignment of a Chemical Hygiene Officer.
9. Requirements for additional protection when working with particularly hazardous substances including "select carcinogens," reproductive toxins, and substances with a high degree of acute toxicity.
10. Provisions for yearly re-evaluation of the CHP.

Other Suggested Elements of a Chemical Hygiene Plan

1. Hazard identification including proper labeling of containers of hazardous chemicals and maintaining MSDSs in a readily accessible location.
2. Requirements to establish and maintain accurate records monitoring employee exposures and any medical consultation and/or examinations, and to assure the confidentiality of these records.

For additional information on developing a CHP consult the following sources:

- *Handbook of Chemical Health and Safety (ACS Handbooks)* by Robert J Alaimo (2001)
- *Prudent Practices in the Laboratory: Handling and Disposal of Chemicals* by The National Research Council (1995)

What is a Material Safety Data Sheet?

Material Safety Data Sheets (MSDS) contains information regarding the proper procedures for handling, storing, and disposing of chemical substances.

- An MSDS accompanies all chemicals or kits that contain chemicals.
- If an MSDS does not accompany a chemical, many web sites and science supply companies can supply one or they can be obtained from www.msdsonline.com.
- Save all MSDSs and store in a designated file or binder using a system that is organized and easy to understand.
- Place the MSDS collection in a central, easily accessible location known to all workers and emergency personnel.
- Typically the information is listed in a standardized format (ANSI Z400.1-1998, Hazardous Industrial Chemicals-Material Safety Data Sheet-Preparation).
- Refer to Appendices I and J for additional information on the format and content of MSDSs (ANSI format).

What Should be Considered When Purchasing Chemicals?

- Establish a chemical procurement plan.
- Consider using a centralized purchasing program in which one person, who is knowledgeable of all the chemicals on hand, does all the purchasing, or links purchasing requests into an inventory tracking system so that excess chemicals in stock can be used before buying more.
- Train receiving room, storeroom, and stockroom personnel in the proper methods of receiving and handling of hazardous substances.

Do the following before ordering chemicals:

- Assess all the hazards and physical properties of the chemical using the MSDS; evaluate both short and long term risks.
- Consider the worst case scenario(s) in the event that the substance is mismanaged, spilled, or causes personal injury.
- Make sure the hazardous properties of the chemical do not exceed the educational utility of the experiment (refer to section titled *Substances with Greater Hazardous Nature than Educational Utility*).
- Determine whether a safer, less hazardous chemical can be used (refer to section titled *What are Some Strategies to Reduce the Amount and/or Toxicity of Chemical Waste Generated in the Laboratory?*).
- Determine whether the appropriate facilities are available for the proper storage of the chemical and the ventilation is sufficient.
- Determine whether the proper personal protective equipment and safety equipment is on hand for using the chemical.
- Establish whether the chemical or its end product will require disposal as a hazardous waste.
- Ensure that the budget will allow for the appropriate and legal disposal of the chemical and/or its end product.
- Have a mechanism in place to dispose of the chemical and its end product legally and safely.
- Determine whether lesser amounts of a chemical can be used to conduct the experiment (refer to section titled *What are Some Strategies to Reduce the Amount and/or Toxicity of Chemical Waste Generated in the Laboratory?*).

When ordering chemicals, remember to do the following:

- Order minimum quantities that are consistent with the rate of use.
- Order only what will be used within a year or less.
- If possible, order reagents in polyethylene bottles or plastic coated glass bottles to minimize breakage, corrosion, and rust.

What is a Chemical Tracking System and How Should It Be Set Up?

A chemical tracking system is a database of chemicals in the laboratory.

A "cradle-to-grave" chemical tracking system should track chemicals from the time they are purchased through the time they are used and discarded.

A good chemical tracking system can reduce procurement costs, eliminate unnecessary purchases, and minimize disposal expenses.

A tracking system can be set up by (1) using index cards or another paper system organized by chemical name and/or molecular formula or (2) by creating a computer-based system.

The following tracking fields are recommended:

- Chemical name as printed on the container
- Chemical name as it appears on the MSDS if different from that on the container
- Molecular formula
- Chemical Abstract Service (CAS) registry number
- Date received
- Source (i.e., chemical manufacturer, and if known, supplier)
- Type of container
- Hazard classification (for storage, handling, and disposal)
- Required storage conditions
- Room number (for larger institutions with multiple storage locations)
- Location within the room (i.e., shelf #1, acid cabinet)
- Expiration or "use by" date
- Amount of the chemical in the container
- Name of the person who ordered or requested the chemical

Each record represents a SINGLE CONTAINER of a chemical (rather than just the chemical itself).

Keep accurate, up-to-date records of the use of each chemical in the system.

Conduct regularly scheduled inventory inspections to purge any inaccurate data in the system and dispose of outdated, unneeded, or deteriorated chemicals following the written Chemical Hygiene Plan.

How Should Chemical Containers Be Labeled?

No unlabeled substance should be present in the laboratory at any time!

Labeling Basics

- Use labels with good adhesive.
- Use a permanent marker (waterproof and fade resistant) or laser (not inkjet) printer.
- Print clearly and visibly.
- Replace damaged, faded or semi-attached labels.

Commercially Packaged Chemicals

Verify that the label contains the following information:
- Chemical name (as it appears on the MSDS)
- Name of chemical manufacturer
- Necessary handling and hazard information

Add:
- Date received
- Date first opened
- Expiration or "use by" date (if one is not present)

Secondary Containers and Prepared Solutions

When one transfers a material from the original manufacturer's container to other vessels, these vessels are referred to as "secondary containers."

Label all containers used for storage with the following:
- Chemical name (as it appears on the MSDS)
- Name of the chemical manufacturer or person who prepared the solution
- Necessary handling and hazard information
- Concentration or purity

- Date prepared
- Expiration or "use by" date

Containers in Immediate Use

These chemicals are to be used within a work shift or laboratory session.

Label all containers in immediate use with the following:

- Chemical name (as it appears on the MSDS)
- Necessary handling and hazard information

Chemical Waste

All containers used for chemical waste should be labeled with:

- "WASTE" or "HAZARDOUS WASTE"
- Chemical name (as it appears on the MSDS)
- Accumulation start date
- Hazard(s) associated with the chemical waste

Peroxide-Forming Substance

Peroxide-forming chemical must be labeled with:

- Date received
- Date first opened
- Date to be disposed of

NOTE: Some States also require (1) National Fire Protection Association (NFPA) code (refer to APPENDIX B) and/or (2) CAS number to be listed on the label. Consult the State regulations.

How Should Chemicals Be Stored?

First, identify any specific requirements regarding the storage of chemicals from (1) local, State, and Federal regulations and (2) insurance carriers.

General Rules for Chemical Storage

Criteria for Storage Area

- Store chemicals inside a closeable cabinet or on a sturdy shelf with a front-edge lip to prevent accidents and chemical spills; a ¾-inch front edge lip is recommended.
- Secure shelving to the wall or floor.
- Ensure that all storage areas have doors with locks.
- Keep chemical storage areas off limits to all students.
- Ventilate storage areas adequately.

Organization

- Organize chemicals first by COMPATIBILITY—not alphabetic succession (refer to section titled *Suggested Shelf Storage Pattern*—next page).
- Store alphabetically within compatible groups.

Chemical Segregation

- Store acids in a dedicated acid cabinet. Nitric acid should be stored alone unless the cabinet provides a separate compartment for nitric acid storage.
- Store highly toxic chemicals in a dedicated, lockable poison cabinet that has been labeled with a highly visible sign.
- Store volatile and odoriferous chemicals in a ventilated cabinet.
- Store flammables in an approved flammable liquid storage cabinet (refer to section titled *Suggested Shelf Storage Pattern*).
- Store water sensitive chemicals in a water-tight cabinet in a cool and dry location segregated from all other chemicals in the laboratory.

Storage Don'ts

- Do not place heavy materials, liquid chemicals, and large containers on high shelves.

- Do not store chemicals on tops of cabinets.
- Do not store chemicals on the floor, even temporarily.
- Do not store items on bench tops and in laboratory chemical hoods, except when in use.
- Do not store chemicals on shelves above eye level.
- Do not store chemicals with food and drink.
- Do not store chemicals in personal staff refrigerators, even temporarily.
- Do not expose stored chemicals to direct heat or sunlight, or highly variable temperatures.

Proper Use of Chemical Storage Containers

- Never use food containers for chemical storage.
- Make sure all containers are properly closed.
- After each use, carefully wipe down the outside of the container with a paper towel before returning it to the storage area. Properly dispose of the paper towel after use.

Suggested Shelf Storage Pattern

A suggested arrangement of compatible chemical families on shelves in a chemical storage room, suggested by the *Flinn Chemical Catalog/Reference Manual*, is depicted on the following page. However, the list of chemicals below does not mean that these chemicals should be used in a high school laboratory.

- First sort chemicals into organic and inorganic classes.
- Next, separate into the following compatible families.

Inorganics	**Organics**
1. Metals, Hydrides	1. Acids, Anhydrides, Peracids
2. Halides, Halogens, Phosphates, Sulfates, Sulfites, Thiosulfates	2. Alcohols, Amides, Amines, Glycols, Imides, Imines
3. Amides, Azides*, Nitrates* (except Ammonium nitrate), Nitrites*, Nitric acid	3. Aldehydes, Esters, Hydrocarbons
4. Carbon, Carbonates, Hydroxides, Oxides, Silicates	4. Ethers*, Ethylene oxide, Halogenated hydrocarbons, Ketenes, Ketones
5. Carbides, Nitrides, Phosphides, Selenides, Sulfides	5. Epoxy compounds, Isocyanates
6. Chlorates, Chlorites, Hydrogen Peroxide*, Hypochlorites, Perchlorates*, Perchloric acid*, Peroxides	6. Azides*, Hydroperoxides, Peroxides
7. Arsenates, Cyanates, Cyanides	7. Nitriles, Polysulfides, Sulfides, Sulfoxides
8. Borates, Chromates, Manganates, Permanganates	8. Cresols, Phenols
9. Acids (except Nitric acid)	
10. Arsenic, Phosphorous*, Phosphorous Pentoxide*, Sulfur	

*Chemicals deserving special attention because of their potential instability.

Suggested Shelf Storage Pattern for Inorganics

ACID STORAGE CABINET ACID INORGANIC #9 Acids, EXCEPT Nitric acid – Store Nitric acid away from other acids unless the cabinet provides a separate compartment for nitric acid storage **Do not store chemicals on the floor**	**Inorganic #10** Arsenic, Phosphorous, Phosphorous Pentoxide, Sulfur	**Inorganic #7** Arsenates, Cyanates, Cyanides STORE AWAY FROM WATER
	Inorganic #2 Halides, Halogens, Phosphates, Sulfates, Sulfites, Thiosulfates	**Inorganic #5** Carbides, Nitrides, Phosphides, Selenides, Sulfides
	Inorganic #3 Amides, Azides, Nitrates, Nitrites EXCEPT Ammonium nitrate - STORE AMMONIUM NITRATE AWAY FROM ALL OTHER SUBSTANCES	**Inorganic #8** Borates, Chromates, Manganates, Permanganates
	Inorganic #1 Hydrides, Metals STORE AWAY FROM WATER. STORE ANY FLAMMABLE SOLIDS IN DEDICATED CABINET	**Inorganic #6** Chlorates, Chlorites, Hypochlorites, Hydrogen Peroxide, Perchlorates, Perchloric acid, Peroxides
	Inorganic #4 Carbon, Carbonates, Hydroxides, Oxides, Silicates	**Miscellaneous**

Suggested Shelf Storage Pattern for Organics

Organic #2 Alcohols, Amides, Amines, Imides, Imines, Glycols **STORE FLAMMABLES IN A DEDICATED CABINET**	**Organic #8** Cresols, Phenol	**POISON STORAGE CABINET** Toxic substances
Organic #3 Aldehydes, esters, hydrocarbons **STORE FLAMMABLES IN A DEDICATED CABINET**	**Organic #6** Azides, Hydroperoxides, Peroxides	**FLAMMABLE STORAGE CABINET** **FLAMMABLE ORGANIC #2** Alcohols, Glycols, etc.
Organic #4 Ethers, Ethylene oxide, Halogenated Hydrocarbons, Ketenes, Ketones **STORE FLAMMABLES IN A DEDICATED CABINET**	**Organic #1** Acids, Anhydrides, Peracids **STORE CERTAIN ORGANIC ACIDS IN ACID CABINET**	**FLAMMABLE ORGANIC #3** Hydrocarbons, Esters, etc.
Organic #5 Epoxy compounds, Isocyanates	Miscellaneous	**FLAMMABLE ORGANIC #4**
Organic #7 Nitriles, Polysulfides, Sulfides, Sulfoxides, etc.	Miscellaneous	**Do not store chemicals on the floor**

How Should Compressed Gas Cylinders Be Stored, Maintained, and Handled?

Compressed gases can be hazardous because each cylinder contains large amounts of energy and may also have high flammability and toxicity potential.

The following is a list of recommendations for storage, maintenance, and handling of compressed gas cylinders:

- Make sure the contents of the compressed gas cylinder are clearly stenciled or stamped on the cylinder or on a durable label.
- Do not identify a gas cylinder by the manufacturer's color code.
- Never use cylinders with missing or unreadable labels.
- Check all cylinders for damage before use.
- Be familiar with the properties and hazards of the gas in the cylinder before using.
- Wear appropriate protective eyewear when handling or using compressed gases.
- Use the proper regulator for each gas cylinder.
- Do not tamper with or attempt to repair a gas cylinder regulator.
- Never lubricate, modify, or force cylinder valves.
- Open valves slowly using only wrenches or tools provided by the cylinder supplier directing the cylinder opening away from people.
- Check for leaks around the valve and handle using a soap solution, "snoop" liquid, or an electronic leak detector.
- Close valves and relieve pressure on cylinder regulators when cylinders are not in use.
- Label empty cylinders "EMPTY" or "MT" and date the tag; treat in the same manner that you would if it were full.
- Always attach valve safety caps when storing or moving cylinders.
- Transport cylinders with an approved cart with a safety chain; never move or roll gas cylinders by hand.
- Securely attach all gas cylinders (empty or full) to a wall or laboratory bench with a clamp or chain, or secure in a metal base in an upright position.
- Store cylinders by gas type, separating oxidizing gases from flammable gases by either 20 feet or a 30-minute firewall that is 5 feet high.

- Store gas cylinders in cool, dry, well-ventilated areas away from incompatible materials and ignition sources.
- Do not subject any part of a cylinder to a temperature higher than 125 °F or below 50 °F.
- Store empty cylinders separately from full cylinders.

What are Some Strategies to Reduce the Amount and/or Toxicity of Chemical Waste Generated in the Laboratory?

All laboratories that use chemicals inevitably produce chemical waste that must be properly disposed of. It is crucial to minimize both the toxicity and the amount of chemical waste that is generated.

A waste management and reduction policy that conforms to State and local regulations should be established by the school or school district.

Several things that can be done to minimize hazards, waste generation, and control costs:

- Purchase chemicals in the smallest quantity needed.
- Use safer chemical substitutes/alternatives such as chemicals which have been determined to be less harmful or toxic (Table 1 contains examples).
- Use microscale experiments.
    - Chemical experiments using smaller quantities of chemicals
- Recycle chemicals by performing cyclic experiments where one product of a reaction becomes the starting material of the following experiment.
- Consider detoxification or waste neutralization steps.
- Use interactive teaching software and demonstration videos in lieu of experiments that generate large amounts of chemical waste.
- Perform classroom demonstrations.
- Use preweighed or premeasured chemical packets such as chemcapsules that reduce bulk chemical disposal problems (no excess chemicals remain).

For information about the EPA's Green Chemistry Program, which promotes the use of innovative technologies to reduce or eliminate the use or generation of hazardous substances, visit:

- www.epa.gov/greenchemistry/
- www.chemistry.org/portal/a/c/s/1/acsdisplay.html?DOC=greenchemistryinstitute/index.html

Table 1. Possible substitutions

Toxic chemicals/equipment	Possible substitution(s)
Mercury thermometers	Digital and alcohol thermometers
Mercury barometer	Aneroid or digital pressure sensors
Methyl orange or methyl red	Bromophenol blue, bromothymol blue
Lead chromate	Copper carbonate
p-Dichlorobenzene	Lauric acid
Dichromate/sulfuric acid mixture	Ordinary detergents, enzymatic cleaners
Alcoholic potassium hydroxide	Ordinary detergents, enzymatic cleaners

What is the Recommended Procedure for Chemical Disposal?

Any chemical discarded or intended to be discarded is chemical waste.

HAZARDOUS chemical waste as designated by the Environmental Protection Agency (EPA) or State authority is waste that presents a danger to human health and/or the environment.

According to EPA regulations, there are four characteristics that define a waste as hazardous:

- Ignitability
- Corrosivity
- Reactivity
- Toxicity

In addition, there are lists of hundreds of other chemicals that EPA has determined to be hazardous waste.

Because of particular differences within some States, consult your State or regional EPA office to determine whether waste is considered hazardous and the requirements for storage and disposal.

For chemical waste, it may be best to use a log book to contain detailed lists of materials in a container labeled "organic waste", for example.

Storing Chemical Waste

- Store all waste in containers that are in good condition and are compatible with their contents.
- Clearly and permanently label each container as to its contents and label as hazardous waste (refer section titled *How Should Chemical Containers Be Labeled?* for specific information).
- Store waste in a designated area away from normal laboratory operations and to prevent unauthorized access.
- Store waste bottles away from sinks and floor drains.
- Do not completely fill waste bottles; leave several inches of space at the top of each waste container.
- Cap all waste bottles.

Proper Disposal of Chemical Waste

The EPA has written a comprehensive set of regulations that govern the management of hazardous waste from the point of generation to ultimate disposal (www.epa.gov/epaoswer/osw/conserve/clusters/schools/index.htm)

Generators of hazardous waste are responsible for ensuring proper disposal of their hazardous waste and can incur liability for improper disposal of their hazardous waste.

Disposal Procedure

- Do not pour chemicals down the drain (unless authorized by local sewer authority).
- Do not treat hazardous waste on-site.
- Contact professional, licensed hazardous waste haulers/transporters that will ensure appropriate disposal.

Appendix A. Common Safety Symbols

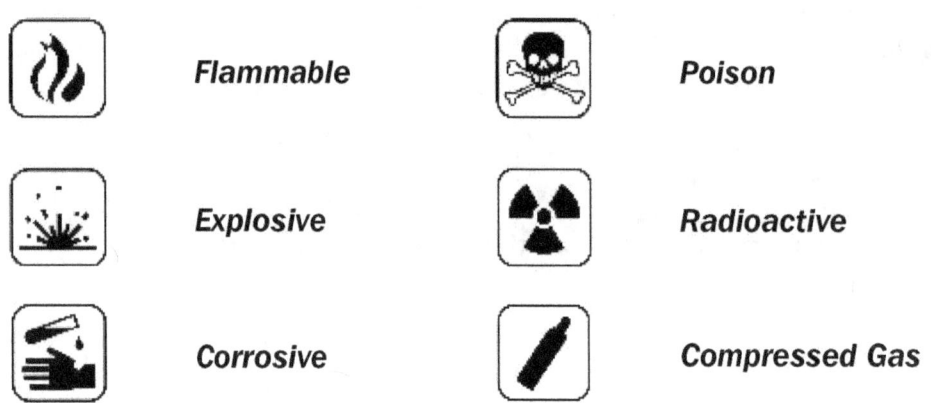

The above safety symbols may be replaced by the following symbols that are internationally accepted*:

Globally Harmonized System of Classification and Labeling of Chemicals, United Nations New York and Geneva, 2005

Appendix B. National Fire Protection Association Hazard Labels

The National Fire Protection Association (NFPA) has developed a visual guide (right) for a number of chemicals pertinent to the MSDS. The ANSI/NFPA 704 Hazard Identification system, the NFPA diamond, is a quick visual review of the health hazard, flammability, reactivity, and special hazards a chemical may present.

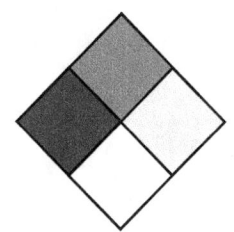

The diamond is broken into four sections (blue, red, yellow, and white). The symbols and numbers in the four sections indicate the degree of hazard associated with a particular chemical or material.

 ### Health Hazard (Blue)

4	Danger	May be fatal on short exposure. Specialized protective equipment required
3	Warning	Corrosive or toxic. Avoid skin contact or inhalation
2	Warning	May be harmful if inhaled or absorbed
1	Caution	May be irritating
0		No unusual hazard

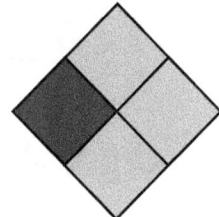

 ### Flammability (Red)

4	Danger	Flammable gas or extremely flammable liquid
3	Warning	Combustible liquid flash point below 100 °F
2	Caution	Combustible liquid flash point of 100° to 200 °F
1		Combustible if heated
0		Not combustible

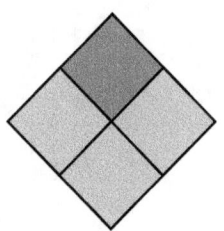

Appendix B: National Fire Protection Association Hazard Labels

 Reactivity (Yellow)

4	Danger	Explosive material at room temperature
3	Danger	May be explosive if shocked, heated under confinement or mixed with water
2	Warning	Unstable or may react violently if mixed with water
1	Caution	May react if heated or mixed with water but not violently
0	Stable	Not reactive when mixed with water

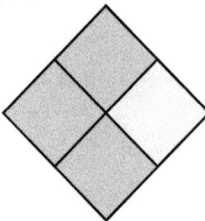

 Special Notice Key (White)

| W | Water Reactive |
| OX | Oxidizing Agent |

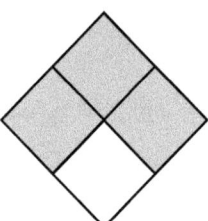

Appendix C. Substances With Greater Hazardous Nature Than Educational Utility

Chemicals used in the laboratory may be hazardous because of the following:

- Safety risks (i.e., highly flammable or explosive material)
- Acute and chronic health hazards
- Environmental harm
- Impairment of indoor air quality

Assessment of the chemicals in this list indicates that their hazardous nature is greater than their potential usefulness in many school programs. Evaluation included physical hazards (i.e., flammability, explosive propensity, reactivity, corrosivity) and health hazards (i.e., toxicity, carcinogenicity).

This following list of chemicals was generated from the *Manual of Safety and Health Hazards in the School Science Laboratory* published by U.S. Department of Health and Human Services, National Institute for Occupational Safety and Health [1984].

Carcinogenic substances were identified from the *Report on Carcinogens* (10th Edition) generated by the National Toxicology Program (2002).

Chemical	CAS Number	Hazard
Acrylonitrile	107–13–1	Flammable (NFPA = 3), reasonably anticipated human carcinogen
Ammonium chromate	7788–98–9	Oxidizer, known human carcinogen
Aniline	62–53–3	Combustible, may be fatal if inhaled, ingested or absorbed through the skin
Aniline hydrochloride	142–04–1	May be fatal if inhaled, ingested, or absorbed through the skin
Anthracene	102–12–7	Irritant, may cause an allergic skin reaction

(Continued)

Appendix C: Substances With Greater Hazardous Nature Than Educational Utility

Chemical	CAS Number	Hazard
Antimony trichloride	10025–91–9	Corrosive
Arsenic and its compounds	N/A	Known human carcinogen
Asbestos	1332–21–4	Known human carcinogen
Ascarite II	N/A	Corrosive, may be fatal if ingested
Benzene	71–43–2	Flammable (NFPA = 3), known human carcinogen, mutagen
Benzoyl peroxide	94–36–0	Flammable (NFPA = 3), explosive, oxidizer
Calcium cyanide	592–01–8	May be fatal if inhaled or ingested
Carbon disulfide	75–15–0	Flammable (NFPA = 4), acute cns toxicity and peripheral neurotoxicity
Carbon tetrachloride	56–23–5	May be fatal if inhaled or ingested, reasonably anticipated human carcinogen
Chloral hydrate	302–17–0	Controlled barbiturate
Chlorine	7782–50–5	Oxidizer, corrosive, may be fatal if inhaled
Chloroform	67–66–3	Reasonably anticipated human carcinogen
Chloropromazine	50–53–3	Controlled substance
Chromium hexavalent compounds	N/A	Known human carcinogen
Chromium trioxide	1333–82–0	Oxidizer, Corrosive, known human carcinogen

(Continued)

Appendix C: Substances With Greater Hazardous Nature Than Educational Utility

Chemical	CAS Number	Hazard
Colchicine	64–86–8	May be fatal if ingested, mutagen
p-Dichlorobenzene	106–46–7	Combustible, reasonably anticipated human carcinogen
Dimethylaniline	121–69–7	May be fatal if inhaled, ingested, or absorbed through the skin
p-Dioxane	123–91–1	Flammable (NFPA = 3), forms peroxides (Group 2), reasonably anticipated human carcinogen
Ethylene dichloride (1,2-Dichloroethane)	107–06–2	Flammable (NFPA = 3), reasonably anticipated human carcinogen, mutagen
Ethylene oxide	75–21–8	Flammable (NFPA = 4), explosive (NPFA = 3), may be fatal if inhaled or absorbed through the skin, known human carcinogen
Gunpowder	N/A	Explosive
Hexachlorophene	70–30–4	May be fatal if inhaled, ingested or absorbed through the skin, possible teratogen
Hydrobromic acid	10035–10–6	Corrosive, may be fatal if inhaled or ingested
Hydrofluoric acid	7664–39–3	Corrosive, may be fatal if inhaled or ingested (liquid and vapor can cause severe burns not always immediately painful or visible but possibly fatal)
Hydrogen	1333–74–0	Flammable (NFPA = 4)
Hydriodic acid	10034–85–2	Corrosive, may be fatal if inhaled or ingested

(Continued)

Appendix C: Substances With Greater Hazardous Nature Than Educational Utility

Chemical	CAS Number	Hazard
Lead arsenate	7784–40–9	Known human carcinogen, teratogen
Lead carbonate	1319–46–6	May be fatal if inhaled or ingested, neurotoxic
Lead (VI) chromate	7758–97–6	May be fatal if inhaled or ingested, known human carcinogen
Lithium, metal	7439–93–2	Combustible, water reactive
Lithium nitrate	7790–69–4	Oxidizer
Magnesium, metal (powder)	7439–95–4	May ignite spontaneiously on contact with water or damp materials
Mercury	7439–97–6	Corrosive, may be fatal if inhaled or ingested
Mercuric chloride	7487–94–7	May be fatal if inhaled, teratogen
Methyl iodide (iodomethane)	74–88–4	May be fatal if inhaled, ingested or absorbed through the skin, potential carcinogen (NIOSH)
Methyl methacrylate	80–62–6	Flammable (NFPA = 3), explosive (vapor)
Methyl orange	547–58–0	Possible mutagen
Methyl red	493–52–7	Possible mutagen
Nickel, metal	7440–02–0	Reasonably anticipated human carcinogen, mutagen
Nickel oxide	1314–06–3	Reasonably anticipated human carcinogen, mutagen
Nicotine	45–11–5	May be fatal if inhaled, ingested, or absorbed through the skin

(Continued)

Appendix C: Substances With Greater Hazardous Nature Than Educational Utility

Chemical	CAS Number	Hazard
Osmium tetroxide	20816–12–0	May be fatal if inhaled or ingested
Paris green	12002–03–8	May be fatal if inhaled, ingested or absorbed through the skin, known human carcinogen
Phenol	108–95–2	Combustible (liquid and vapor), corrosive, may be fatal if inhaled, ingested or absorbed through the skin
Phosphorus pentoxide	1314–56–3	Water reactive, corrosive
Phosphorous, red, white	7723–14–0	May ignite spontaneously in air
Phthalic anhydride	85–44–9	Combustible/finely dispersed particles form explosive mixtures in air, corrosive
Potassium, metal	7440–09–7	Flammable (nfpa = 3), water reactive, forms peroxides
Potassium oxalate	583–52–8	Corrosive, may be fatal if ingested
Potassium sulfide	1312–73–8	Spontaneously combustible, explosive in dust or powder form, corrosive
Pyridine	110–86–1	Flammable (nfpa = 3), possible mutagen
Selenium	7782–49–2	Severe irritant
Silver cyanide	506–64–9	May be fatal if inhaled, ingested or absorbed through the skin
Silver nitrate	7761–88–8	Oxidizer, corrosive, may be fatal if ingested
Silver oxide	20667–12–3	Oxidizer

(Continued)

Appendix C: Substances With Greater Hazardous Nature Than Educational Utility

Chemical	CAS Number	Hazard
Sodium arsenate	7778–43–0	May be fatal if inhaled or ingested, known human carcinogen
Sodium arsenite	7784–46–5	Known human carcinogen, teratogen
Sodium azide	26628–22–8	Explosive, may be fatal if ingested or absorbed through the skin
Sodium chromate	7775–11–3	Oxidizer, corrosive, known human carcinogen
Sodium cyanide	143–33–9	May be fatal if inhaled, ingested or absorbed through the skin
Sodium dichromate	10588–01–9	Oxidizer, corrosive, may be fatal if ingested, known human carcinogen
Sodium nitrite	7632–00–0	Oxidizer
Sodium sulfide	1313–82–2	Corrosive, may be fatal if inhaled or ingested
Sodium thiocyanide	540–72–7	Contact with acid liberates very toxic gas
Stannic chloride (anhydrous)	7646–78–8	Corrosive, hydrochloric acid liberated upon contact with moisture and heat
Stearic acid	57–11–4	May form combustible dust concentration in the air
Strontium	7440–24–6	Water reactive
Strontium nitrate	10042–76–9	Oxidizer

(Continued)

Appendix C: Substances With Greater Hazardous Nature Than Educational Utility

Chemical	CAS Number	Hazard
Sudan IV	85–83–6	Irritant, toxic properties have not been thoroughly evaluated
Sulfuric acid, fuming	8014–95–7	Corrosive, may be fatal if ingested
Tannic acid	1401–55–4	Irritant
Tetrabromoethane	79–27–6	May be fatal if inhaled, ingested or absorbed through the skin
Thioacetamide	62–55–5	Reasonably anticipated human carcinogen
Thiourea	62–56–6	Reasonably anticipated human carcinogen
Titanium trichloride	7705–07–9	Water reactive, corrosive
Titanium tetrachloride	7550–45–0	Water reactive, corrosive, may be fatal if inhaled
o-Toluidine	95–53–4	Reasonably anticipated human carcinogen, mutagen
Uranium	7440–61–1	Radioactive material
Uranyl acetate	541–09–3	Radioactive material
Urethane	51–79–6	Combustible, reasonably anticipated human carcinogen
Wood's metal	8049–22–7	May be fatal if inhaled or ingested, known human carcinogen (cadmium), neurotoxic

Appendix D. Substances With a Hazardous Nature, but May Have Potential Educational Utility

These chemicals should be removed from the schools if alternatives can be used. For those that must be retained, amounts should be kept to a minimum. These are appropriate for advanced-level High School classes only.

This following list was generated from the *Manual of Safety and Health Hazards in the School Science Laboratory* published by U.S. Department of Health and Human Services, National Institute for Occupational Safety and Health [1984].

Carcinogenic substances were identified from the *Report on Carcinogens* (10th Edition) generated by the National Toxicology Program (2002).

Chemical	CAS Number	Hazard
Acetamide	60–35–5	Combustible solid
Aluminum chloride	7446–70–0	Water reactive, corrosive
Ammonium bichromate	7789–09–5	Oxidizer, corrosive, known human carcinogen
Ammonium oxalate	1113–38–8	May be fatal if inhaled or ingested
Ammonium vanadate	7803–55–6	May be fatal if inhaled or ingested
Antimony	7440–36–0	May be fatal if inhaled, irritant
Antimony oxide	1309–64–4	Irritant
Antimony potassium tartrate	11071–15–1	Irritant
Barium chloride	10361–37–2	May be fatal if ingested, irritant

(Continued)

Appendix D: Substances With a Hazardous Nature, but May Have Potential Educational Utility

Chemical	CAS Number	Hazard
Benzone (phenylbutazone)	50–33–9	Irritant
Beryllium carbonate	66104–24–3	Irritant
Bromine	7726–95–6	Oxidizer, corrosive, may be fatal if inhaled or ingested
Cadmium and cadmium compounds	N/A	Known human carcinogen
Carmine	860–22–0	Irritant, burning may produce carbon monoxide, carbon dioxide, sulfur oxides, and nitrogen oxides.
Catechol	120–80–9	Corrosive
Chromic acid	7738–94–5	Oxidizer, known human carcinogen
Chromium acetate	1066–30–4	Irritant
Cobalt, metal	7440–48–4	Possible human carcinogen (IARC, Group 2B)
Cobalt nitrate	10141–05–6	Oxidizer, irritant
Cyclohexane	110–82–7	Flammable (NFPA = 3)
Cyclohexene	110–83–8	Flammable (nfpa = 3), corrosive, forms peroxides
Dichloroindophenol sodium salt	620–45–1	Irritant
2,4-Dinitrophenol	51–28–5	Irritant
Ferrous Sulfate	7720–78–7	Irritant
Formaldehyde (formalin)	50–00–0	Flammable (NFPA = 3), reasonably anticipated human carcinogen

(Continued)

Appendix D: Substances With a Hazardous Nature, but May Have Potential Educational Utility

Chemical	CAS Number	Hazard
Fuchsin (acid/basic)	3244–88–0/ 632–99–5	Irritant
Gasoline	8006–61–9	Flammable (NFPA = 3)
Hematoxylin	517–28–2	Irritant
Hydrogen sulfide	7783–06–4	Corrosive
Hydroquinone	123–31–9	May be fatal if ingested
Isoamyl alcohol (isopentyl alcohol)	123–51–3	Irritant, combustible liquid and vapor
Isobutyl alcohol	78–83–1	Flammable (NFPA = 3)
Magnesium chlorate	10326–21–3	Irritant
Methyl ethyl ketone	78–93–3	Irritant, flammable (NFPA = 3)
Methyl oleate	112–62–9	Toxic properties not investigated
Nickel carbonate	3333–67–3	Reasonably anticipated human carcinogen
Nickelous acetate	373–02–4	Reasonably anticipated human carcinogen
Paradichlorobenzene	106–46–7	Irritant
Pentane	109–66–0	Irritant, flammable (NFPA = 4)
Petroleum ether	8032–32–4	Flammable (NFPA = 4)
1-Phenyl-2-Thiourea (Phenylthiocarbamide)	103–85–5	May be fatal if inhaled or ingested
Potassium chlorate	3811–04–9.	Oxidizer

(Continued)

Appendix D: Substances With a Hazardous Nature, but May Have Potential Educational Utility

Chemical	CAS Number	Hazard
Potassium chromate	7789–00–6	Oxidizer, known human carcinogen
Potassium periodate	7790–21–8	Oxidizer
Potassium permanganate	7722–64–7	Oxidizer, corrosive
Salol (phenyl salicylate)	118–55–8	Irritant
Sodium bromate	7789–38–0	Oxidizer
Sodium chlorate	7775–09–9.	Oxidizer
Sodium fluoride	7681–49–4	May be fatal if inhaled or ingested
Sodium oxalate	62–76–0	Corrosive, may be fatal if ingested
Sodium nitrate	7631–99–4	Oxidizer, irritant
Sodium silicofluoride	16893–85–9	Toxic
Sudan III	85–86–9	Decomposes to oxides of nitrogen
Sulfamethazine	57–68–1	Irritant
Toluene	108–88–3	Flammable (NFPA = 3), irritant, may be fatal if ingested
Trichloroethylene	79–01–6	Reasonably anticipated human carcinogen
Urethane	51–79–6	Combustible, reasonably anticipated human carcinogen
Xylenes	1330–20–7	Flammable (NFPA = 3), irritant, may be fatal if ingested

Appendix E. Incompatible Chemicals

This list represents the commonly used laboratory chemicals and their incompatibilities with other chemicals. This list was generated from the *Hazards in the Chemical Laboratory*, 4th Edition, Safety in Academic CL. Bretherick, Ed. [1986]; reproduced by permission of the Royal Society of Chemistry. It is by no means complete; however, it can be used as a guide for proper storage and use in the laboratory. Specific incompatibilities are also listed in the material safety data sheets.

Chemical	Incompatible with
Acetic acid	Chromic acid, Nitric acid, Peroxides, Permanganates
Acetic anhydride	Hydroxyl group containing compounds, Ethylene glycol, Perchloric acid)
Acetone	Concentrated Nitric and Sulfuric acid mixtures, Hydrogen peroxide
Acetylene	Bromine, Chlorine, Copper, Fluorine, Mercury, Silver
Ammonium nitrate	Acids, Chlorates, Flammable liquids, Nitrates, powdered metals, Sulphur, finely divided organic or combustible materials
Aniline	Hydrogen peroxide, Nitric acid
Calcium oxide	Water
Carbon, activated	Calcium hypochlorite, other oxidants
Chlorates	Acids, Ammonium salts, Metal powders, Sulphur, finely divided organic or combustible materials

(Continued)

Appendix E: Incompatible Chemicals

Chemical	Incompatible with
Chromic acid	Acetic acid, Camphor, Glycerol, Naphthalene, Turpentine, other flammable liquids
Chlorine	Acetylene, Ammonia, Benzene, Butadiene, Butane and other petroleum gases, Hydrogen, Sodium carbide, Turpentine, finely divided metals,
Copper	Acetylene, Hydrogen peroxide
Hydrazine	Hydrogen peroxide, Nitric acid, other oxidants
Hydrocarbons	Bromine, Chlorine, Chromic acid, Fluorine, peroxides
Hydrocyanic acid	Alkalis, Nitric acid
Hydrofluoric acid, anhydrous	Ammonia (aqueous or anhydrous)
Hydrogen peroxide	Aniline, Chromium, combustible materials, Copper, Iron, most metals and their salts, Nitromethane, any flammable liquid
Hydrogen sulfide	Fuming nitric acid, oxidizing gases
Iodine	Acetylene, Ammonia (aqueous or anhydrous)
Mercury	Acetylene, Ammonia, Fulminic acid
Nitric acid, concentrated	Acetic acid, Acetone, Alcohol, Aniline, Chromic acid, flammable gases, flammable liquids, Hydrocyanic acid, Hydrogen Sulfide, Nitratable substances
Nitroparaffins	Amines, inorganic bases
Oxalic acid	Mercury, Silver
Oxygen	Flammable liquids, solids, or gases, grease, Hydrogen, oils
Perchloric acid	Acetic anhydride, Alcohol, Bismuth and its alloys, grease, oils, paper, wood

(Continued)

Appendix E: Incompatible Chemicals

Chemical	Incompatible with
Peroxides, organic	Acids (organic or mineral)
Phosphorus (white)	Air, Oxygen
Potassium chlorate	Acids (also refer to chlorates)
Potassium perchlorate	Acids (also refer to percholoric acid)
Potassium permanganate	Benzaldehyde, Ethylene glycol, Glycerol, Sulfuric acid
Silver	Acetylene, Ammonium compounds, Fulminic acid, Oxalic acid, Tartaric acid,
Sodium	Carbon dioxide, Carbon tetrachloride and other chlorinated compounds, water
Sodium nitrite	Ammonium nitrate and other ammonium salts
Sodium peroxide	Any oxidizable substances (e.g., Acetic anhydride, Benzaldehyde, Carbon disulfide, Ethanol, Ethyl acetate, Ethylene glycol, Furfural, Glacial acetic acid, Methanol, Methyl acetate)
Sulphuric acid	Chlorates, Perchlorates, Permanganates

Appendix F. Recommended Safety and Emergency Equipment for the Laboratory

The following are checklists for safety and emergency equipment for the laboratory:

Personal Protective Equipment

- ✔ Chemical splash goggles
- ✔ Face shields
- ✔ Lab coat
- ✔ Lab apron
- ✔ Gloves (selected based on the material being handled and the particular hazard involved)

Safety and Emergency Equipment

- ✔ Hand-free eye-wash stations (not eye-wash bottles) that conform to ANSI Z358.1–2004
- ✔ Deluge safety showers that conform to ANSI Z358.1–2004
- ✔ Safety shields with heavy base
- ✔ Fire extinguishers (dry chemical and carbon dioxide extinguishers)
- ✔ Sand bucket
- ✔ Fire blankets
- ✔ Emergency lights
- ✔ Emergency signs and placards
- ✔ Fire detection or alarm system with pull stations
- ✔ First-aid kits
- ✔ Spill control kit (absorbent and neutralizing agents)
- ✔ Chemical storage cabinets (preferably with an explosion proof ventilation system)
- ✔ Gallon-size carrying buckets for chemical bottles
- ✔ Laboratory chemical hood (60–100 ft/minute capture velocity, vented outside)
- ✔ Ground-fault interrupter electrical outlets
- ✔ Container for broken glass and sharps
- ✔ Material Safety Data Sheets (MSDSs)
- ✔ Emergency Action Plan for the institution

Appendix G. How Does a Chemical Enter the Body?

- A chemical can enter the body through different routes.
- These different routes of exposure and the types of exposure (acute or chronic) can affect the toxicity of the chemical.
- The most probable (primary) route(s) of exposure to a chemical will be identified in the MSDS.
- Three principal routes of exposure include: dermal exposure (skin), inhalation, and ingestion (oral).

Dermal Exposure

Although the skin is an effective barrier for many chemicals, it is a common route of exposure. The toxicity of a chemical depends on the degree of absorption that occurs once it penetrates the skin. Once the skin is penetrated, the chemical enters the blood stream and is carried to all parts of the body. Chemicals are absorbed much more readily through injured, chapped, or cracked skin, or needle sticks than through intact skin. Generally, organic chemicals are much more likely to penetrate the skin than inorganic chemicals.

Dermal exposure to various substances can also cause irritation and damage to the skin and/or eyes. Depending on the substance and length of exposure, effects of dermal exposures can range from mild temporary discomfort to permanent damage.

Inhalation

Inhalation is another route of chemical exposure. Chemicals in the form of gases, vapors, mists, fumes, and dusts entering through the nose or mouth can be absorbed through the mucous membranes of the nose, trachea, bronchi, and lungs. Unlike the skin, lung tissue is not a very protective barrier against the access of chemicals into the body. Chemicals, especially organic chemicals, enter into the blood stream quickly. Chemicals can also damage the lung surface.

Ingestion

Ingestion involves chemicals entering the body through the mouth. Chemical dusts, particles and mists may be inhaled through the mouth and swallowed.

They may also enter through contaminated objects, such as hands or food that come in contact with the mouth. Absorption of the chemicals into the bloodstream can occur anywhere along the length of the gastrointestinal (GI) tract.

Appendix H. What Are Exposure Limits?

Exposure limits are intended to protect workers from excessive exposure to hazardous substances:

- Established by health and safety authorities and chemical manufacturers
    - Department of Labor's Occupational Safety and Health Administration (OSHA)
    - American Conference of Governmental Industrial Hygienists (ACGIH)
    - National Institute for Occupational Safety and Health (NIOSH)
    - Environmental Protection Agency (EPA)
    - American Industrial Hygiene Association (AIHA)
- Define the amount/concentration to which a worker can be exposed without causing an adverse health effect.
- Typically pertain to the concentration of a chemical in the air, but may also define limits for physical agents such as noise, radiation, and heat.
- Usually can be found on the MSDS; make sure your MSDSs are up-to-date.

Exposure Limits

Legally Enforceable Limits

Permissible Exposure Limits (PELs)

- Set by OSHA, 29 CFR 1910.1000, and 1910.1001 through 1910.1450.
- Specifies the maximum amount or concentration of a chemical to which a worker may be exposed.
- Generally defined in three different ways
    1. **Ceiling Limit (C):** the concentration that must not be exceeded at any part of the workday
    2. **Short Term Exposure Limit (STEL):** the maximum concentration to which workers may be exposed for a short period of time (15 minutes)
    3. **Time Weighted Average (TWA):** the average concentration to which workers may be exposed for a normal, 8-hour workday

Other U.S. Exposure Limits

Threshold Limit Values (TLVs)

- Prepared by ACGIH volunteer scientists
- Denotes the level of exposure that nearly all workers can experience without an unreasonable risk of disease or injury
- An advisory limit; not enforceable by law
- Generally can be defined as ceiling limits, short term exposure limits, and/or time-weighted averages
- Usually equivalent to PELs

Recommended Exposure Limits (RELs)

- Recommended by NIOSH
- Indicates the concentration of a substance to which a worker can be exposed for up to a 10-hour workday during a 40-hour work week without adverse effects, however, sometimes based on technical feasibility
- Based on animal and human studies
- Generally expressed as a ceiling limit, short-term exposure limit, or a time-weighted average
- Often more conservative than PELs and TLVs

Workplace Environmental Exposure Limits (WEELs)

- Developed by AIHA volunteers
- Advisory limits; not enforceable by law
- Typically developed for chemicals that are not widely used or for which little toxicity information is available

Company-Developed Limits

- Developed by company scientists
- Advisory limits; not enforceable by law
- Usually based on only short-term studies of animals
- Generally intended for internal company use and sometimes for the customers

Appendix I. General Guidelines to Follow in the Event of a Chemical Accident or Spill

- Assess the overall situation.
- Determine the appropriate action to resolve the situation.
- Follow the pre-existing, approved local emergency plan.
- Act swiftly and decisively.

Below are some recommended actions for specific emergencies. Some of the actions have been proposed by the Council of State Science Supervisors in Science & Safety: Making the Connection.

Chemical in the Eye

- Flush the eye immediately with water while holding the eye open with fingers.
- If wearing contact lens, remove and continue to rinse the eye with water.
- Continue to flush the eye and seek immediate medical attention.

Acid/Base Spill

For a spill not directly on human skin, do the following:

- Neutralize acids with powdered sodium hydrogen carbonate (sodium bicarbonate/baking soda), or bases with vinegar (5% acetic acid solution).
- Avoid inhaling vapors.
- Spread diatomaceous earth to absorb the neutralized chemical.
- Sweep up and dispose of as hazardous waste.

For spills directly on human skin, do the following:

- Flush area with copious amounts of cold water from the faucet or drench shower for at least 5 minutes.
- If spill is on clothing, first remove clothing from the skin and soak the area with water as soon as possible.
- Arrange treatment by medical personnel.

Mercury Spill

- Evacuate the affected area.

Appendix I: General Guidelines to Follow in the Event of a Chemical Accident or Spill

- Close off interior doors and windows, and heating and air conditioning vents in the incident room.
- Open exterior doors and windows to move the inside air outside.
- Follow specific cleanup instructions detailed by the EPA (www.epa.gov/epaoswer/hazwaste/mercury/spills.htm) or by your state.

Appendix J. Understanding an MSDS

ANSI Standardized MSDS Format

Section 1 gives details on *what the chemical or substance is, CAS number, synonyms*, the *name of the company* issuing the data sheet, and often an *emergency contact number*.

Section 2 identifies the *OSHA hazardous ingredients*, and may include *other key ingredients* and exposure limits.

Section 3 lists the *major health effects* associated with the chemical. Sometimes both the acute and chronic hazards are given.

Section 4 provides *first aid measures* that should be initiated in case of exposure.

Section 5 presents the *fire-fighting measures* to be taken.

Section 6 details the *procedures to be taken in case of an accidental release*. The instructions given may not be sufficiently comprehensive in all cases, and local rules and procedures should be utilized to supplement the information given in the MSDS sheet.

Section 7 addresses the *storage and handling* information for the chemical. This is an important section as it contains information on the flammability, explosive risk, propensity to form peroxides, and chemical incompatibility for the substance. It also addresses any special storage requirements for the chemical (i.e., special cabinets or refrigerators).

Section 8 outlines the *regulatory limits for exposure*, usually the maximum permissible exposure limits (PEL) (refer to Appendix G). The PEL, issued by the Occupational Safety and Health Administration, tells the concentration of air contamination a person can be exposed to for 8 hours a day, 40 hours per week over a working lifetime (30 years) without suffering adverse health effects. It also provides information on personal protective equipment.

Section 9 gives the *physical and chemical properties* of the chemical. Information such as the evaporation rate, specific gravity, and flash points are given.

Section 10 gives the *stability and reactivity* of the chemical with information about chemical incompatibilities and conditions to avoid.

Section 11 provides both the *acute and chronic toxicity* of the chemical and any health effects that may be attributed to the chemical.

Section 12 identifies both the *ecotoxicity* and the environmental fate of the chemical.

Section 13 offers suggestions for the *disposal of the chemical*. Local, state, and Federal regulations should be followed.

Section 14 gives the *transportation information* required by the Department of Transportation. This often identifies the dangers associated with the chemical, such as flammability, toxicity, radioactivity, and reactivity.

Section 15 outlines the *regulatory information* for the chemical. The hazard codes for the chemical are given along with principle hazards associated with the chemical. A variety of country and/or state specific details may be given.

Section 16 provides *additional information* such as the label warnings, preparation and revision dates, name of the person or firm that prepared the MSDS, disclaimers, and references used to prepare the MSDS.

Appendix K. Sample MSDS

Material Safety Data Sheet

Toluene MSDS No. XXXX

1. Product and Company Identification

Product Name: TOLUENE
Synonyms: Methylbenzene, Methylbenzol, Phenylmethane, Toluol
CAS No.: 108–88–3
Chemical Formula: C6H5–CH3
Catalog Number: Tol 12

Supplier: Company X
XXXXXXXXX
Anywhere, XX XXXXX

Emergency Information: 800–XXX–XXXX

2. Composition/Information on Ingredients

Ingredient	CAS No	Percent	Hazardous
Toluene	108–88–3	100%	Yes

3. Hazards Identification

Emergency Overview

DANGER! Harmful or fatal if swallowed. Vapor harmful. **POISON!** May be absorbed through intact skin. Flammable liquid and vapor. May cause liver and kidney damage, may affect blood system or central nervous system. Causes irritation to skin, eyes and respiratory tract.

Potential Acute Health Effects

- **Eye Contact:** Causes severe eye irritation with redness and pain.
- **Skin Contact:** Causes irritation. May be absorbed through skin.
- **Inhalation:** Inhalation may cause irritation of the upper respiratory tract. Symptoms of overexposure may include fatigue, confusion, headache, dizziness and drowsiness. Very high concentrations may cause unconsciousness and death.

- **Ingestion:** Swallowing may cause abdominal spasms and other symptoms that parallel over-exposure from inhalation. Aspiration of material into the lungs may cause chemical pneumonitis, which may be fatal.
- **Chronic Exposure:** Chronic exposure may result in anemia, decreased blood cell count and bone marrow hypoplasia. Liver and kidney damage may occur. Repeated or prolonged contact may cause dermatitis.

4. First Aid Measures

Eye Contact: Immediately flush eyes with plenty of water for at least 15 minutes, lifting the upper and lower eye lids occasionally. Get medical attention immediately.

Skin Contact: In case of contact, immediately flush skin with plenty of soap and water for at least 15 minutes while removing contaminated clothing and shoes. Wash clothing before reuse. Call a physician immediately.

Inhalation: Evacuate victim to fresh air immediately. If not breathing, give artificial respiration. If breathing is difficult, give oxygen. Seek medical aid immediately.

Ingestion: Aspiration hazard. If swallowed, DO NOT INDUCE VOMITING. Give 2–4 cups of milk or water. Never give anything by mouth to an unconscious person. Get medical attention immediately.

5. Fire Fighting Measures

Fire: Flash point: 4 °C (40 °F)
Autoignition temperature: 480 °C (896 °F)
Flammable limits in air % by volume: lower: 1.3%; upper: 7.1%
Flammable liquid and vapor!
Extremely flammable when exposed to flame or sparks. Vapors are heavier than air and can flow along surfaces to distant ignition source and flash back.

Explosion: Vapor-air concentrations above flammable limits are explosive. Contact with strong oxidizers may cause fire or explosion. Sensitive to static discharge.

Fire Extinguishing Media: Dry chemical, carbon dioxide or foam. Material is lighter than water and a fire may be spread by use of water. Water may be used to cool fire surface and protect personnel. Water may also be used to flush spills away from exposures and to dilute spills to non-flammable mixtures. Avoid flushing hydrocarbon into sewers.

Appendix K: Sample MSDS

Special Information: In the event of a fire, wear full protective clothing and NIOSH-approved self-contained breathing apparatus operated in the pressure demand or other positive pressure mode.

6. Accidental Release Measures

Avoid contact: Ventilate area of leak or spill. Remove all ignition sources. Wear appropriate personal protective equipment as specified in Section 8. Isolate hazard area. Contain and recover liquid when possible. Collect liquid in an appropriate container or absorb with an inert material such as earth, sand or vermiculite. Do not use combustible materials, such as saw dust. Do not flush to sewer.

7. Handling and Storage

Handling: Wash thoroughly after handling. Use with adequate ventilation. Avoid contact with skin, eyes or clothes. Electrically ground and bond containers when transferring material to avoid static accumulation.

Storage: Store in a cool, dry well-ventilated location, away from any area where the fire hazard. Separate from incompatibles. Storage and use areas should be No Smoking areas. Use non-sparking type tools and equipment, including explosion proof ventilation. Containers of this material may be hazardous when empty since they retain product residues (vapors, liquid). Observe all warnings and precautions listed for the product. Protect container against physical damage. Keep container tightly closed.

8. Exposure Controls/Personal Protection

Ventilation System: A system of local and/or general exhaust is recommended to keep exposures below the Airborne Exposure Limits.

Exposure Limits: Toluene:

- OSHA Permissible Exposure Limit (PEL): 200 ppm TWA; 300 ppm (acceptable ceiling conc.); 500 ppm (acceptable maximum conc.).
- NIOSH Recommended Exposure Limit (REL): 100 ppm TWA (375 mg/m^3); STEL 150 ppm (560 mg/m^3)
- ACGIH Threshold Limit Value (TLV): 50 ppm TWA skin – potential for cutaneous absorption

Personal Respirators (NIOSH/EN 149 Approved): If the exposure limit is exceeded a half-face organic vapor respirator may be worn for up to ten times the exposure limit. A full-face organic vapor respirator or self-contained breathing apparatus may be worn up to 50 times the exposure limit. For emergencies or instances where the exposure levels are not known, use a full-face piece positive-pressure, air-supplied respirator.

Skin Protection: Wear impervious protective clothing, including boots, gloves, lab coat, apron or coveralls, as appropriate, to prevent skin contact.

Eye Protection: Use chemical splash goggles and/or a full face shield. Maintain eyewash fountain facilities in work area.

9. Physical and Chemical Properties

Physical State and appearance: Clear, colorless liquid.
Odor: Aromatic benzene-like.
Solubility: Very slight
Specific Gravity (Water = 1): 0.9
Viscosity: 20cP @ 20 °C
Boiling Point: 110 °C (232 °F)
Melting Point: −95 °C (−139 °F)
Vapor Density (Air=1): 3.1
Vapor Pressure (mm Hg): 53.3 @ 20 °C (68 °F)
Evaporation Rate (Butyl acetate=1): 2.4
Molecular formula: $C_6H_5CH_3$
Molecular weight: 92.06

10. Stability and Reactivity

Stability: Stable under ordinary conditions of use and storage. Containers may burst when heated.

Hazardous Decomposition Products: Carbon dioxide and carbon monoxide may form when heated to decomposition.

Hazardous Polymerization: Has not been reported.

Incompatibilities: Heat, flame, strong oxidizers, nitric and sulfuric acids; will attack some forms of plastics, rubber, coatings.

Conditions to Avoid: Heat, flames, ignition sources and incompatibles.

11. Toxicological Information

Toxicological Data:

Oral rat LD_{50}: 636 mg/kg
Inhalation rat LC_{50}: 49 gm/m³/4H

Skin rabbit LD_{50}: 14100 uL/kg
Inhalation mouse LC_{50}:
400 ppm/24H

Irritation data: skin rabbit, 500 mg, Eye rabbit, 2 mg/24H, Severe.
Moderate

Investigated as a tumorigen, mutagen, reproductive effector.

Reproductive Toxicity:
Has shown some evidence of reproductive effects in laboratory animals.

12. Ecological Information

Environmental Fate: When released into the soil, this material may evaporate and is microbiologically biodegradable. When released into the soil, this material is expected to leach into groundwater. When released into water, this material may evaporate and biodegrade to a moderate extent. When released into the air, this material may be moderately degraded by reaction with photochemically produced hydroxyl radicals.

Environmental Toxicity: No data available, however this material is expected to be toxic to aquatic life.

13. Disposal Considerations

Waste material should be handled as hazardous waste and sent to a RCRA approved incinerator or disposed in a RCRA approved waste facility. Processing, use or contamination of this product may change the waste management options. State and local disposal regulations may differ from Federal disposal regulations. Dispose of container and unused contents in accordance with Federal, State and local requirements.

14. Transport Information

Domestic (Land, U.S. D.O.T.)

Proper Shipping Name: TOLUENE
Hazard Class: 3
UN/NA: UN1294
Packing Group: II

Canada TDG

Proper Shipping Name: TOLUENE
Hazard Class: 3 (9.2)
UN/NA: UN1294
Packing Group: II
Additional Information: Flashpoint 4 °C

15. Regulatory Information

CALIFORNIA PROPOSITION 65: WARNING

This product contains a chemical known to the State of California to cause birth defects or other reproductive harm.
Reportable Quantity: 1000 Pounds (454 Kilograms) (138.50 Gals)

NFPA Rating: Health – 2; Fire – 3; Reactivity – 0
0=Insignificant 1=Slight 2=Moderate 3=High 4=Extreme

Carcinogenicity Lists: No
NTP: No
IARC Monograph: No
OSHA Regulated: No

Section 313 Supplier Notification: This product contains the following toxic chemical(s) subject to the reporting requirements of SARA TITLE III Section 313 of the Emergency Planning and Community Right-To-Know Act of 1986 and of 40 CFR 372:

CAS No.	Chemical Name	% By Weight
108–88–3	Toluene	100

16. Other Information

Label Hazard Warning

POISON! DANGER! HARMFUL OR FATAL IF SWALLOWED. HARMFUL IF INHALED OR ABSORBED THROUGH SKIN. VAPOR HARMFUL. FLAMMABLE LIQUID AND VAPOR. MAY AFFECT LIVER, KIDNEYS, BLOOD SYSTEM, OR CENTRAL NERVOUS SYSTEM. CAUSES IRRITATION TO SKIN, EYES AND RESPIRATORY TRACT.

Label Precautions

Keep away from heat, sparks and flame.
Keep container closed.
Use only with adequate ventilation.
Wash thoroughly after handling.
Avoid breathing vapor.
Avoid contact with eyes, skin and clothing.

Label First Aid

Aspiration hazard. If swallowed, DO NOT INDUCE VOMITING. Give large quantities of water. Never give anything by mouth to an unconscious person. If vomiting occurs, keep head below hips to prevent aspiration into lungs. If inhaled, remove to fresh air. If not breathing, give artificial respiration. If breathing is difficult, give oxygen. In case of contact, immediately flush eyes or skin with plenty of water for at least 15 minutes. Remove contaminated clothing and shoes. Wash clothing before reuse. In all cases call a physician immediately.

References: Upon request

Appendix L. Web Site Resources

Federal Government

Department of Health and Human Services
Centers for Disease Control and Prevention (CDC)
www.cdc.gov

Consumer Product Safety Commission (CPSC)
www.cpsc.gov

Department of Transportation (DOT)
www.dot.gov

Environmental Protection Agency (EPA)
www.epa.gov

Schools Chemical Cleanout Campaign (SC3)
www.epa.gov/epaoswer/osw/conserve/clusters/schools/index.htm

Department of Health and Human Services
Centers for Disease Control and Prevention
National Institute for Occupational Safety and Health (NIOSH)
www.cdc.gov/niosh/homepage.html

Department of Health and Human Services
National Toxicology Program (NTP)
http://ntp-server.niehs.nih.gov

Department of Labor
Occupational Safety and Health Administration (OSHA)
www.osha.gov

Other

American Chemical Society (ACS)
www.acs.org

Council of State Science Supervisors (CSSS)
www.csss-science.org/safety.htm

International Agency for Research on Cancer (IARC)
www.iarc.fr

Laboratory Safety Institute (LSI)
www.labsafety.org

MSDS Online
www.msdsonline.com

National Fire Protection Association (NFPA)
www.nfpa.org

National Safety Council (NSC)
www.nsc.org

National Science Teachers Association (NSTA)
www.nsta.org

Safety Information Resources Inc (SIRI) MSDS Collection
www.hazard.com

Appendix M. Glossary

Acid
A substance that dissolves in water and releases hydrogen ions (H+); acids cause irritation, burns, or more serious damage to tissue, depending on the strength of the acid, which is measured by pH.

Acute toxicity
Adverse effects resulting from a single dose, or exposure to a substance for less than 24 hours.

Allergy
An exaggerated immune response to a foreign substance causing tissue inflammation and organ dysfunction.

Asphyxiant
A substance that interferes with the transport of an adequate supply of oxygen to the body by either displacing oxygen from the air or combining with hemoglobin, thereby reducing the blood's ability to transport oxygen.

Base
A substance that dissolves in water and releases hydroxide ions (OH–); bases cause irritation, burns, or more serious damage to tissue, depending on the strength of the base, which is measured by pH.

Carcinogen
A substance that causes cancer.

CAS Registry number
An internationally recognized unique registration number assigned by the Chemical Abstracts Service to a chemical, a group of similar chemicals, or a mixture.

Ceiling limit
The maximum permissible concentration of a material in the working environment that should never be exceeded for any duration.

Chemical hygiene plan
A written program that outlines procedures, equipment, and work practices that protect employees from the health hazards present in the workplace.

Chemical hygiene officer
A designated person who provides technical guidance in the development and implementation of the Chemical Hygiene Plan.

Chronic toxicity
Adverse effects resulting from repeated doses of, or exposures to, a substance by any route for more than three months.

Combustible liquid
A liquid with a flashpoint at a temperature lower than the boiling point; according to the National Fire Protection Association and the U.S. Department of Transportation, it is a liquid with a flash point of 100 °F (37.8 °C) or higher.

Compatible materials
Substances that do not react together to cause a fire, explosion, violent reaction or lead to the evolution of flammable gases or otherwise lead to injury to people or danger to property.

Compressed gas
A substance in a container with an absolute pressure greater than 276 kilopascals (kPa) or 40 pounds per square inch (psi) at 21 °C, or an absolute pressure greater than 717 kPa (40 psi) at 54 °C.

Consumer Product Safety Commission (CPSC)
An independent U.S. Federal regulatory agency that protects the public against unreasonable risk of injury and death associated with consumer products.

Corrosive
A substance capable of causing visible destruction of, and/or irreversible changes to living tissue by chemical action at the site of contact (i.e., strong acids, strong bases, dehydrating agents, and oxidizing agents).

Department of Transportation (DOT)
U.S. Federal agency that regulates the labeling and transportation of hazardous materials.

Environmental Protection Agency (EPA)
U.S. Federal agency that develops and enforces regulations to protect human health and the natural environment.

Explosive
A substance that causes a sudden, almost instantaneous release of pressure, gas, and heat when subjected to sudden shock, pressure, or high temperature.

Exposure limits
The concentration of a substance in the workplace to which most workers can be exposed during a normal daily and weekly work schedule without adverse effects.

Federal Hazardous Substances Act (FHSA)
The Federal Hazardous Substances Act (15 U.S.C 1261–1278), administered by the Consumer Product Safety Commission, requires that certain household products that are "hazardous substances" bear cautionary labeling to alert consumers to potential hazards that those products present and inform them of the measures they need to protect themselves from those hazards. Any product that is toxic, corrosive, flammable or combustible, an irritant, a strong sensitizer, or that generates pressure through decomposition, heat, or other means requires labeling, if the product may cause substantial personal injury or substantial illness during or as a proximate result of any customary or reasonable foreseeable handling or use, including reasonable foreseeable ingestion by children.

Flammable
As defined in the FHSA regulations at 16 CFR § 1500.3(c)(6)(ii), a substance having a flashpoint above 20 °F (−6.7 °C) and below 100 °F (37.8 °C). An extremely flammable substance, as defined in the FHSA regulations at 16 CFR § 1500.3(c)(6)(i), is any substance with a flashpoint at or below 20oF (−6.7 °C).

Flashpoint
The minimum temperature at which a liquid or a solid produces a vapor near its surface sufficient to form an ignitable mixture with the air; the lower the flash point, the easier it is to ignite the material.

Hazardous substance
As defined in the Federal Hazardous Substances Act (FHSA) at 16 CFR § 1500.3(b)(4)(i)(A), any substance or mixture of substances that is toxic, corrosive, an irritant, a strong sensitizer, flammable or combustible, or generates pressure through decomposition, heat or other means, if it may cause substantial personal injury or illness during or as a proximate result of any customary or reasonably foreseeable handling or use, including reasonably foreseeable ingestion by children.

Hepatotoxin
A chemical that can cause liver damage.

Highly toxic substance
As defined by OSHA (Appendix A of 29 CFR 1910.1200) and in the FHSA regulations at 16 CFR § 1500.3(b)(6)(i), a substance with either (a) a median lethal dose (LD_{50}) of 50 mg/kg or less of body weight administered orally to rats, (b) a median lethal dose (LD_{50}) of 200 mg/kg or less of body weight when

administered continuously on the bare skin of rabbits for 24 hours or less, or (c) a median lethal concentration (LC_{50}) in air of 200 parts per million by volume or less of gas or vapor, or 2 mg/L by volume or less of mist or dust, when administered by continuous inhalation for one hour or less to rats.

Ignitable
Capable of bursting into flames; ignitable substances pose a fire hazard

International Agency for Research on Cancer (IARC)
An agency of the World Health Organization that publishes IARC *Monographs on the Evaluation of the Carcinogenic Risk of Chemicals to Humans*. This publication documents reviews of information on chemicals and determinations of the cancer risk of chemicals.

Incompatible materials
Substances that can react to cause a fire, explosion, violent reaction or lead to the evolution of flammable gases or otherwise lead to injury to people or danger to property.

Ingestion
Taking a substance into the body by mouth and swallowing it.

Inhalation
Breathing a substance into the lungs; substance may be in the form of a gas, fume, mist, vapor, dust, or aerosol.

Irritant
A substance that causes a reversible inflammatory effect on living tissue by chemical action at the site of contact.

Known human carcinogen
A substance for which there is sufficient evidence of a cause and effect relationship between exposure to the material and cancer in humans.

Lacrimation
Excessive production of tears when the eye is exposed to an irritant.

LC_{50} (Median Lethal Concentration 50)
The concentration of a chemical that kills 50% of a sample population; typically expressed in mass per unit volume of air.

LD_{50} (Median Lethal Dose 50)
The amount of a chemical that kills 50% of a sample population; typically expressed as milligrams per kilogram of body weight.

Mutagen
A substance capable of changing genetic material in a cell.

National Fire Protection Association (NFPA)
An organization that provides information about fire protection and prevention and developed a standard outlining a hazard-warning labeling system that rates the hazard(s) of a material during a fire (health, flammability, and reactivity hazards).

National Institute for Occupational Safety and Health (NIOSH)
U.S. Federal agency of the Centers for Disease Control and Prevention (CDC) that investigates and evaluates potential hazards in the workplace. NIOSH is also responsible for conducting research and providing recommendations for the prevention of work-related illness and injuries.

National Toxicology Program (NTP)
U.S. Federal interagency program that coordinates toxicological testing programs, develops and validates improved testing methods, and provides toxicological evaluations on substances of public health concern.

Neurotoxin
A substance that induces an adverse effect on the structure and/or function of the central and/or peripheral nervous system.

Occupational Safety and Health Administration (OSHA)
U.S. Federal agency that develops and enforces occupational safety and health standards for all general, as well as, construction and maritime industries and businesses in the U.S.

Oxidizer
A substance that causes the ignition of combustible materials without an external source of ignition; oxidizers can produce oxygen, and therefore support combustion in an oxygen free atmosphere.

Peroxide former
A substance that reacts with air or oxygen to form explosive peroxy compounds that are shock, pressure, or heat sensitive.

Permissible Exposure Limit (PELs)
The legally enforceable maximum amount or concentration of a chemical that a worker may be exposed to under OSHA regulations.

Personal Protective Equipment (PPE)
Any clothing and/or equipment used to protect the head, torso, arms, hands, and feet from exposure to chemical, physical, or thermal hazards.

pH
A measure of the acidity or basicity (alkalinity) of a material when dissolved in water; expressed on a scale from 0 to 14.

Radioactive material
A material whose nuclei spontaneously give off nuclear radiation.

Reactivity
The capacity of a substance to combine chemically with other substances.

Reproductive toxicity
Adverse effects on sexual function and fertility in adult males and females, as well as developmental toxicity in the offspring (International Programme on Chemical Safety (IPCS) Environmental Health Criteria 225, *Principles for Evaluating Health Risks to Reproduction Associated with Exposure to Chemicals*).

Secondary containment
An empty chemical-resistant container/dike placed under or around chemical storage containers for the purpose of containing a spill should the chemical container leak.

Short Term Exposure Limit (STEL)
The maximum concentration to which workers can be exposed for a short period of time (15 minutes).

Systemic
Affecting many or all body systems or organs; not localized in one spot or area.

Teratogen
A substance which may cause non-heritable genetic mutations or malformations in the developing embryo or fetus when a pregnant female is exposed to the substance.

Threshold Limit Value (TLV)
Term used by the American Conference of Governmental Industrial Hygienists (ACGIH) to express the recommended exposure limits of a chemical to which nearly all workers may be repeatedly exposed, day after day, without adverse effect.

Time Weighted Averages (TWA)
The average concentration to which an average worker can be exposed for a normal, 8 hour workday.

Toxic substance

In general, as defined in the FHSA regulations at 16 CFR § 1500.3(b)(5), any substance (other than a radioactive substance) which has the capacity to produce personal injury or illness to man through ingestion, inhalation, or absorption through any surface of the body.

This term is further defined by OSHA and in the FHSA regulations:

As defined by OSHA (Appendix A of 29 CFR 1910.1200), a substance with either, a median lethal dose (LD_{50}) of more than 50 mg/kg but not more than 500 mg/kg of body weight administered orally, a median lethal dose (LD_{50}) of more than 200 mg/kg but not more than 1,000 mg/kg of body weight when administered by continuous contact with the bare skin of rabbits, or a median lethal concentration (LC_{50}) in air of more than 200 parts per million but not more than 2,000 parts per million by volume of gas or vapor, or more than 2 mg/L but not more than 20 mg/L of mist, fume, or dust, when administered by continuous inhalation for one hour.

As defined in the FHSA regulations at 16 CFR § 1500.3(c)(2)(i), a substance with either, a median lethal dose (LD_{50}) of 50 mg/kg to 5,000 mg/kg of body weight administered orally in rats, a median lethal dose (LD_{50}) of more than 200 mg/kg but not more than 2,000 mg/kg of body weight when administered by continuous contact with the bare skin of rabbits for 24 hours, or a median lethal concentration (LC_{50}) in air of more than 200 parts per million but not more than 20,000 parts per million by volume of gas or vapor, or more than 2 mg/L but not more than 200 mg/L by volume of mist or dust, when administered by continuous inhalation for one hour or less.

Water reactive material

A substance that reacts with water that could generate enough heat for the item to spontaneously combust or explode. The reaction may also release a gas that is either flammable or presents a health hazard.

SAFE LAB

SUPERVISION	Never work in the lab without the supervision of a teacher
ATTENTION	Always pay attention to the work—don't fool around in the lab
FOLLOW INSTRUCTIONS	Always perform experiments precisely as directed by the teacher
EMERGENCY PREPAREDNESS	Know what to do in the event of an emergency
LABELING	Check labels to verify substances before using them. Label Containers
APPAREL	Always wear appropriate protective equipment and apparel
BRAINS	Use them—Safety begins with you

SAFETY DO'S AND DON'TS FOR STUDENTS

How Should Chemicals Be Stored?

First, identify any specific requirements regarding the storage of chemicals from (1) local, State, and Federal regulations and (2) insurance carriers.

General Rules for Chemical Storage

Criteria for Storage Area

- Store chemicals inside a closeable cabinet or on a sturdy shelf with a front-edge lip to prevent accidents and chemical spills; a ¾-inch front edge lip is recommended.
- Secure shelving to the wall or floor.
- Ensure that all storage areas have doors with locks.
- Keep chemical storage areas off limits to all students.
- Ventilate storage areas adequately.

Organization

- Organize chemicals first by COMPATIBILITY—not alphabetic succession (refer to section titled *Suggested Shelf Storage Pattern*—next page).
- Store alphabetically within compatible groups.

Chemical Segregation

- Store acids in a dedicated acid cabinet. Nitric acid should be stored alone unless the cabinet provides a separate compartment for nitric acid storage.
- Store highly toxic chemicals in a dedicated, lockable poison cabinet that has been labeled with a highly visible sign.
- Store volatile and odoriferous chemicals in a ventilated cabinet.
- Store flammables in an approved flammable liquid storage cabinet (refer to section titled *Suggested Shelf Storage Pattern*).
- Store water sensitive chemicals in a water-tight cabinet in a cool and dry location segregated from all other chemicals in the laboratory.

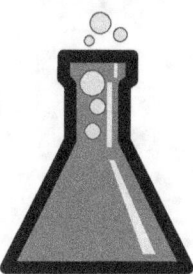

Storage Don'ts

- Do not place heavy materials, liquid chemicals, and large containers on high shelves.
- Do not store chemicals on tops of cabinets.
- Do not store chemicals on the floor, even temporarily.
- Do not store items on bench tops and in laboratory chemical hoods, except when in use.
- Do not store chemicals on shelves above eye level.
- Do not store chemicals with food and drink.
- Do not store chemicals in personal staff refrigerators, even temporarily.
- Do not expose stored chemicals to direct heat or sunlight, or highly variable temperatures.

Proper Use of Chemical Storage Containers

- Never use food containers for chemical storage.
- Make sure all containers are properly closed.
- After each use, carefully wipe down the outside of the container with a paper towel before returning it to the storage area. Properly dispose of the paper towel after use.

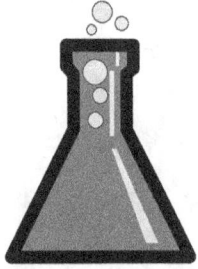

Suggested Shelf Storage Pattern

A suggested arrangement of compatible chemical families on shelves in a chemical storage room, suggested by the *Flinn Chemical Catalog/Reference Manual*, is depicted on the following page. However, the list of chemicals below does not mean that these chemicals should be used in a high school laboratory.

- First sort chemicals into organic and inorganic classes.
- Next, separate into the following compatible families.

Inorganics	**Organics**
1. Metals, Hydrides	1. Acids, Anhydrides, Peracids
2. Halides, Halogens, Phosphates, Sulfates, Sulfites, Thiosulfates	2. Alcohols, Amides, Amines, Glycols, Imides, Imines
3. Amides, Azides*, Nitrates* (except Ammonium nitrate), Nitrites*, Nitric acid	3. Aldehydes, Esters, Hydrocarbons
4. Carbon, Carbonates, Hydroxides, Oxides, Silicates	4. Ethers*, Ethylene oxide, Halogenated hydrocarbons, Ketenes, Ketones
5. Carbides, Nitrides, Phosphides, Selenides, Sulfides	5. Epoxy compounds, Isocyanates
6. Chlorates, Chlorites, Hydrogen Peroxide*, Hypochlorites, Perchlorates*, Perchloric acid*, Peroxides	6. Azides*, Hydroperoxides, Peroxides
7. Arsenates, Cyanates, Cyanides	7. Nitriles, Polysulfides, Sulfides, Sulfoxides
8. Borates, Chromates, Manganates, Permanganates	8. Cresols, Phenols
9. Acids (except Nitric acid)	
10. Arsenic, Phosphorous*, Phosphorous Pentoxide*, Sulfur	

*Chemicals deserving special attention because of their potential instability.

Suggested Shelf Storage Pattern for Inorganics

SAFE LAB

**ACID STORAGE CABINET
ACID
INORGANIC #9**

Acids, EXCEPT Nitric acid – Store Nitric acid away from other acids unless the cabinet provides a separate compartment for nitric acid storage

Do not store chemicals on the floor

Inorganic #10 Arsenic, Phosphorous, Phosphorous Pentoxide, Sulfur	**Inorganic #7** Arsenates, Cyanates, Cyanides **STORE AWAY FROM WATER**
Inorganic #2 Halides, Halogens, Phosphates, Sulfates, Sulfites, Thiosulfates	**Inorganic #5** Carbides, Nitrides, Phosphides, Selenides, Sulfides
Inorganic #3 Amides, Azides, Nitrates, Nitrites **EXCEPT Ammonium nitrate - STORE AMMONIUM NITRATE AWAY FROM ALL OTHER SUBSTANCES**	**Inorganic #8** Borates, Chromates, Manganates, Permanganates
Inorganic #1 Hydrides, Metals **STORE AWAY FROM WATER.
STORE ANY FLAMMABLE SOLIDS IN DEDICATED CABINET**	**Inorganic #6** Chlorates, Chlorites, Hypochlorites, Hydrogen Peroxide, Perchlorates, Perchloric acid, Peroxides
Inorganic #4 Carbon, Carbonates, Hydroxides, Oxides, Silicates	**Miscellaneous**

Suggested Shelf Storage Pattern for Organics

SAFE LAB

Organic #2 Alcohols, Amides, Amines, Imides, Imines, Glycols **STORE FLAMMABLES IN A DEDICATED CABINET**	**Organic #8** Cresols, Phenol
Organic #3 Aldehydes, esters, hydrocarbons **STORE FLAMMABLES IN A DEDICATED CABINET**	**Organic #6** Azides, Hydroperoxides, Peroxides
Organic #4 Ethers, Ethylene oxide, Halogenated Hydrocarbons, Ketenes, Ketones **STORE FLAMMABLES IN A DEDICATED CABINET**	**Organic #1** Acids, Anhydrides, Peracids **STORE CERTAIN ORGANIC ACIDS IN ACID CABINET**
Organic #5 Epoxy compounds, Isocyanates	Miscellaneous
Organic #7 Nitriles, Polysulfides, Sulfides, Sulfoxides, etc.	Miscellaneous

POISON STORAGE CABINET

Toxic substances

FLAMMABLE STORAGE CABINET

FLAMMABLE ORGANIC #2

Alcohols, Glycols, etc.

FLAMMABLE ORGANIC #3

Hydrocarbons, Esters, etc.

FLAMMABLE ORGANIC #4

Do not store chemicals on the floor

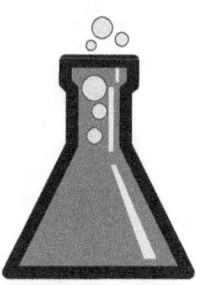

www.ingramcontent.com/pod-product-compliance
Lightning Source LLC
Chambersburg PA
CBHW081734170526
45167CB00009B/3816